Apples 🍎🍎 for Teachers

A Basic Skills Reinforcement Program for Young Children

Numbers 0-10

by
Diane Burkle
Cynthia Polk Muller
and
Linda J. Petuch

Fearon Teacher Aids
Belmont, California

Illustrator: Marilynn G. Barr

ISBN 0-8224-0457-5

Printed in the United States of America

1. 9 8 7 6 5 4

contents

introduction

Numbers 0—10 is designed to enhance primary education through the use of manipulatives and tactile materials. The learning activities help students develop visual, auditory, kinesthetic, tactile, and perceptual skills while providing opportunities for creative expression. Although students may appear to be participating in art projects, they are, in fact, achieving artistically such curriculum objectives as recognizing numbers, associating numbers and numerals, and developing eye-hand coordination and fine motor skills.

The worksheets contained here are structured to allow repetition when learning new concepts—an approach that educators have noted is essential for internalizing concepts and skills. The learning materials are specifically designed to appeal to young children and to provide them with the positive reinforcement they need for learning new skills.

Worksheet-based activities require minimal teacher preparation and can be used with students working individually, in small groups, or in independent learning centers. Most of these activities use common school supplies such as crayons, scissors, and glue or paste. Some activities require additional materials that are inexpensive and readily available in many classrooms or from art supply stores.

To administer an activity, assemble and distribute the necessary materials. (Materials lists, along with suggestions for variations, extensions, and enrichment activities, are provided in the section called "Using the Worksheets.") Read the worksheet instructions aloud to the students and make sure they understand what to do. Provide help as needed and always supervise the use of scissors and glue.

We hope you enjoy using this book and, most of all, that it makes learning fun for your students.

using the worksheets

There are, on the average, seven worksheets to teach each number, 0 through 10. The worksheets are grouped by number, with review worksheets inserted where appropriate. After your students have completed one or two worksheets of a kind, they may be able to complete the others of that kind without instructions. Use the following notes to help you administer each kind of worksheet.

learning the number

(pp. 13, 17, 24, 31, 38, 45, 56, 63, 70, 77, and 84)

objectives and skills

Recognizing number shapes
Developing left-to-right orientation
Developing fine motor skills

variations

✓ For activities that require glue, have students use glue sticks instead. Glue sticks, available at most stationery stores or art supply stores, are convenient and keep artwork (and children's hands!) neat and clean. When children use glue sticks to fill in the number outlines, they may actually be simulating number-writing.

✓ You may have children fill in the outline numbers with any available materials, or you may have them repeat the crayon-resist activity (described for numbers 0, 5, and 10) for all other numbers.

✓ Since the concept "0" is difficult for most young children, you may wish to introduce this number after students have learned numbers 1 through 5.

extensions

✓ Have students cut out the numbers and glue them onto contrasting background, such as colored construction paper.

✓ Create a bulletin board of cutout numbers, allowing students to sort and sequence the numbers.

✓ Have the children act out each number as they learn it, forming number shapes with their bodies.

materials

Learning the Number 0 (p. 13)
black crayons
water colors
water
small plastic containers for
 diluted paint
paintbrushes

Learning the Number 1 (p. 17)
construction paper in various
 colors
glue

Learning the Number 2 (p. 24)
glue
dry macaroni

Learning the Number 3 (p. 31)
glue
19-inch lengths of thick, colorful
 yarn

Learning the Number 4 (p. 38)
glue
tissue paper in various colors

Learning the Number 5 (p. 45)
black crayons
water colors
water
small plastic containers for
 diluted paint
paintbrushes

Learning the Number 6 (p. 56)
glue
dried beans or peas

Learning the Number 7 (p. 63)
glue
sand

Learning the Number 8 (p. 70)
old newspapers and magazines
scissors
glue

Learning the Number 9 (p. 77)
glue
popped popcorn

Learning the Number 10 (p. 84)
black crayons
water colors
water
small plastic containers for
 diluted paint
paintbrushes

writing the number

(pp. 14, 18, 25, 32, 39, 46, 57, 64, 71, 78, and 85)

objectives and skills

Identifying numbers
Associating numbers of objects with numbers
Developing left-to-right orientation
Developing fine motor skills

materials

crayons

variation

✓You may wish to introduce writing "0" after children have
successfully completed writing "5." Zero is a difficult concept for
many young children.

extension	✓When students have successfully completed pages of this type, they may wish to gather their pages into a *Numbers* booklet. Show students how to bind their booklets with construction paper covers, and help them staple the pages together in order.
enrichment	✓Have students "paint" numbers on the sidewalk or outside walls of the school using large paintbrushes and water.
	✓Ask students to arrange objects on a table to show given numbers of objects.

counting objects (pp. 15, 19, 26, 33, 40, 47, 58, 65, 72, 79, and 86)

objectives and skills	Identifying numbers Associating numbers of objects with numbers Developing eye-hand coordination Developing fine motor skills
materials	crayons scissors glue
variations	✓Have students discard the leftover objects they cut out, instead of gluing these to the back of the page.
	✓Have students draw the correct number of objects instead of cutting and pasting the objects pictured.
extensions	✓Ask the children to write the number of objects beside the picture.
	✓Create a bulletin board display of pictures, grouped by number of objects. The students can help sort the pictures by number and sequence them.
	✓Have students count the objects they have glued to the back of the page.
enrichment	✓Point to various objects in the classroom or outside, and ask students how many there are of those particular objects. For example, how many legs are there on this chair? How many branches are on that tree? How many books are on this shelf?

drawing objects   (pp. 16, 20, 27, 34, 41, 48, 59, 66, 73, 80, and 87)

objectives and skills   Drawing the given number of objects
Associating numbers of objects with numbers
Developing fine motor skills

materials   crayons

variations   ✓ Suggest the students draw objects of a like nature rather than duplicates of the objects depicted. For example, they might draw other types of flowers than the flower pictured, or they might draw other vehicles (buses, bicycles, trucks, or vans) instead of drawing cars like the one pictured.

✓ If students have difficulty drawing small objects, have them use separate sheets of paper for each different item on a page. Or suggest they draw some objects on the back of the page.

✓ The students might also draw simple shapes (circles, squares, and triangles) if they have difficulty drawing the objects pictured.

enrichment   ✓ Ask the children to draw objects that have a given number of features: a bicycle (two wheels), a triangle (three sides, three points), an animal that has four legs, a star (five points), a snowflake (six points), a spider (eight legs). Together with your students, count the features on each object.

learning the value of numerals   (pp. 21, 28, 35, 42, 49, 60, 67, 74, 81, and 88)

objectives and skills   Distinguishing between numbers of objects in groups
Associating numbers of objects with numbers
Developing fine motor skills
Developing eye-hand coordination

materials	Learning the Value of Numeral 1 (p. 21)	Learning the Value of Numeral 5 (p. 49)
	scissors	crayons
	glue	scissors
		glue
	Learning the Value of Numeral 2 (p. 28)	Learning the Value of Numeral 6 (p. 60)
	scissors	crayons
	glue	scissors
		glue
	Learning the Value of Numeral 3 (p. 35)	Learning the Value of Numeral 7 (or 8, 9, or 10) (pp. 67, 74, 81, and 88)
	crayons	crayons
	scissors	
	glue	
	Learning the Value of Numeral 4 (p. 42)	
	crayons	

learning number words

(pp. 16, 22, 29, 36, 43, 50, 61, 68, 75, 82, and 89)

objectives and skills

Recognizing number words
Recognizing numbers
Associating numbers, number words, and numbers of objects
Developing left-to-right orientation

materials

crayons

extension

✓ Provide writing paper for additional practice writing the number words.

enrichment

✓ Write the number words *zero* through *ten* on separate index cards. Gather counting objects (small plastic animals, blocks, paper clips, and so on) on a table. Ask one small group of students to arrange the objects in groups of one to ten. Then have a companion group place the number word cards correctly beside the groups of items.

addition readiness

(pp. 23, 30, 37, 44, 51, 62, 69, 76, and 83)

objectives and skills

Counting objects
Drawing like objects
Recalling numbers

materials   crayons

variation   ✓ Have students write the number word instead of the number to represent the number of objects.

extension   ✓ Capable students may write the number to represent the group of objects on the page, then the number to represent the objects they have drawn (always one), and then the number of total objects.

enrichment   ✓ Conduct the activity using manipulatives such as blocks, small plastic animals, crayons, erasers, and so on. Have the students count out one to nine objects and then add one object to the pile. Then ask them to count the objects in the pile again. If the students are able to count to ten easily, then they will eventually be able to predict what the count will be each time they add an object to the pile.

✓ Try the activity again, using either manipulatives (as described above) or the worksheet, having students add two objects, three objects, or zero objects. (Some students may even be able to count beyond ten.)

reviewing numbers   (pp. 52, 54–55, 90, 93–94, 95)

objectives and skills   Recognizing numbers
Associating numbers of objects with numbers
Distinguishing between numbers of objects in groups
Developing fine motor skills

variations   ✓ Have students color the worksheets before they begin gluing items in place.

✓ For the worksheet on page 52, use styrofoam packing material (often referred to as "popcorn") instead of real popcorn. You should be able to obtain this material from any major department store. Ask someone in the wrapping department.

enrichment   ✓ Construct an interactive bulletin board based on any of the review pages. For example, draw a tree with several branches and 11 empty nests. Label the nests 0 through 10. Attach a shallow envelope inside each nest. From construction paper, cut out about 60 eggs or about 60 small birds, and place them in a pocket at the side of the bulletin board. Have the students slip the correct number of eggs or birds into each shallow envelope placed over the nest. Students may check each other's work.

materials	**Reviewing 0 Through 5** (p. 52)	**Reviewing Groups of 6 to 10**
	popped popcorn (each student needs at least 15 pieces)	(pp. 93–94)
	glue	scissors
		glue
		crayons
	Reviewing Groups of 0 to 5	
	(pp. 54–55)	**Reviewing 0 Through 10**
	scissors	(p. 95)
	glue	small, dried white beans
	crayons	glue
	Reviewing 6 Through 10 (p. 90)	
	sequins (each student needs at least 40 sequins)	
	glue	

graphing numbers (pp. 53 and 91–92)

objectives and skills
Recognizing numbers
Associating numbers of objects with numbers
Developing fine motor skills
Developing eye-hand coordination

materials
scissors
glue

variations
✓ Before the students begin gluing shapes into the columns, show them how a graph looks. Explain that they should start at the baseline (near the number) and stack the shapes up in the column.

✓ Students might color the shapes before they cut them out and glue them.

✓ Instead of using the cutout shapes, provide bright and interesting adhesive stickers for the students to use.

enrichment
✓ Have students work in small groups to create their own graphs of objects in the classroom or around the school grounds. They may be able to use the object to represent itself (gluing real leaves to the graph, for example), or they may draw pictures of the objects. Encourage them to use a different object for each number on the graph.

setting up learning centers

The worksheet-based activities in this book may be adapted to a learning center environment. Nearly every worksheet focuses on a separate number, and the worksheets are usually repeated (sometimes with slight variation) for every number. Hence, even young children who cannot yet read may be able to work at a learning center, without teacher direction, completing worksheets on their own or in small groups. Here are some suggestions that may help you in setting up learning centers in your classroom:

1 Set up a separate learning center for each subject area. Identify the center by hanging a symbol, such as a number, over the table set up for that learning center.

2 Gather materials that students will need for the activity. Place those materials in boxes or other appropriate containers, and label the containers with the word and a picture of the item contained. Store these containers in the learning center.

3 For activities requiring paint, glue, or paste, tape a protective covering over the learning center table. Keep the covering on the table as long as that learning center continues to be used.

4 Prepare samples of the activities. Post the samples on the bulletin board or in the learning center so that students can refer to them.

5 Explain each different activity before the students begin working on their own. You may need to review the directions on a daily basis.

6 If you have a few learning centers set up, students may work at the centers in rotation. Group students by ability, by compatibility, or in some random fashion. Create a chart that shows who is in each group, and post the chart on a bulletin board. Create tags that match the learning center symbols (see suggestion 1), and pin a different tag next to each group listed on the chart. The tag will tell the group which center they should go to that day. Change the tags each day.

The art activities are especially appropriate for cooperative learning situations. You may wish to assign roles, such as gathering materials, cleaning up after the activity, collecting the worksheets and making sure group members have written their names on their papers, and so on. You might also want to provide an appropriate social goal such as working quietly, asking group members for help, and so on.

7 Provide adequate space for the completed worksheets to dry, if they have just been glued, pasted, or painted. Try not to stack or overlap the worksheets while they dry.

Name _____

Color the **0** using a black crayon.
Paint the whole page, brushing from left to right.
Let the paint dry.
Trace the number **0** with your finger.

Name _____

Trace the **0**s.

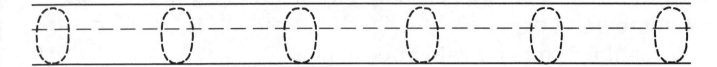

Write some **0**s.

_____
- - - - - - - - - - - - - - - -
_____

Color **0** objects below.

Numbers 0—10 © 1988

Name _____

Color the apple and the worms. Cut out the worms.
Glue **0** worms to the apple.
Glue the other worms onto the back of this page.

Name _____

Trace the number word.

———————————————

- - - zero - - -

———————————————

Write the number word.

———————————————

- - - - - - - - - - -

———————————————

Draw a line from the number word to the correct number.

zero

0 1 2 3 4 5

Draw a line from the number word to the correct number of objects.

zero

Numbers 0—10 © 1988

Name _____

Tear colored paper into small pieces.
Glue the pieces on the number **I**.
Let the glue dry.
Trace the number **I** with your finger.

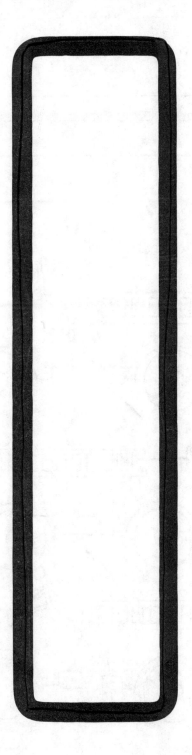

Name _____

Trace the **I**s.

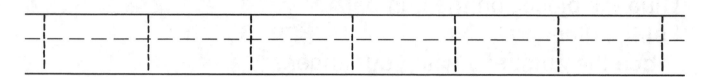

Write some **I**s.

_____

- - - - - - - - - - - - - - - -

_____

Color **I** object below.

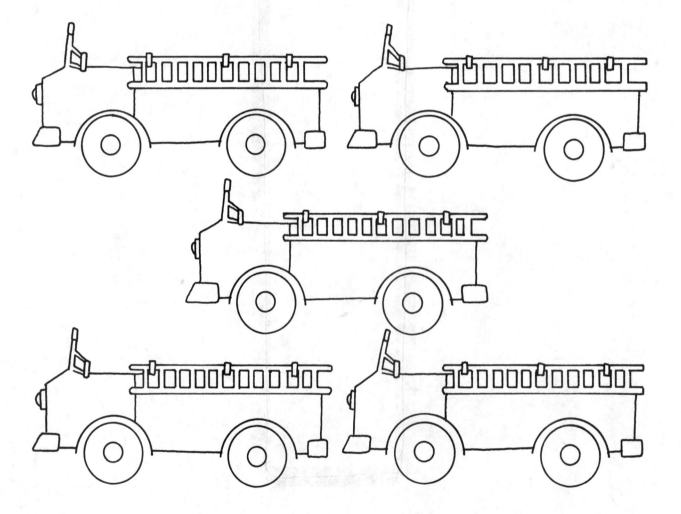

Numbers 0—10 © 1988

Name _____

Color the fishbowl and the fish. Cut out the fish.
Glue **1** fish inside the fishbowl.
Glue the other fish onto the back of this page.

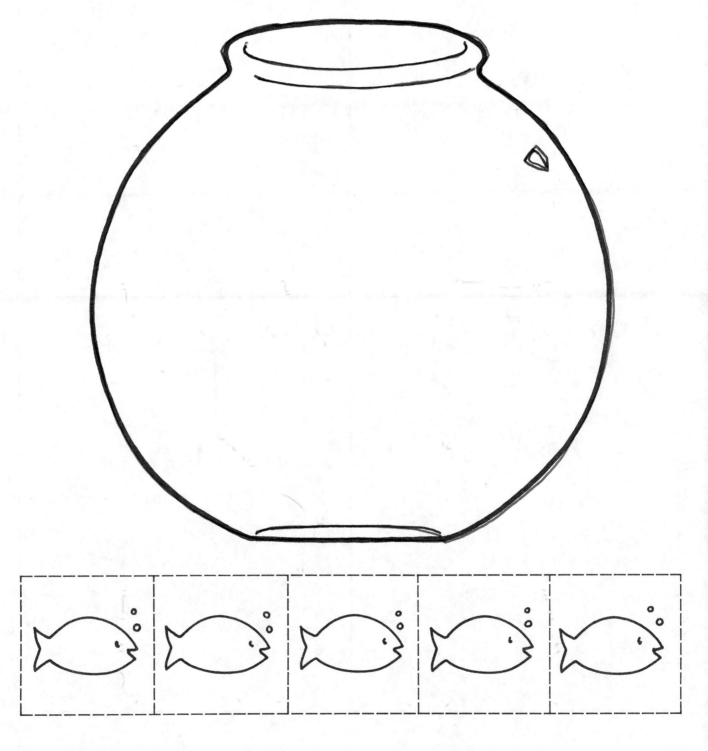

Name _____

Draw **1**

Draw **1**

Draw **1**

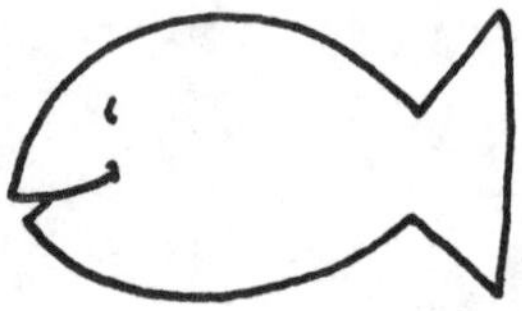

Draw **1**

Name _____

Count the flowers in each square below.
Cut out each square that has **1** flower in it.
Glue those squares inside the flower vase.

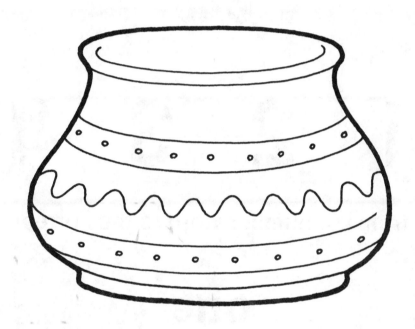

Write the number **1**.

_____

- - - - - - - - - - - - - - - - - - - -

_____

Name _____

Trace the number word.　　　Write the number word.

_____　　_____

- - - - -ᴏ̣n̤ᴇ̣- - - - - -　　- - - - - - - - - - -

Draw a line from the number word to the correct number.

one

0 I 2 3 4 5

Draw a line from the number word to the correct number of objects.

one

　　Numbers 0—10 © 1988

Name _____

Here is a carton of milk.

Draw one more carton of milk.

Now how many cartons of milk are there?
Write the number.

_____
– – – – – – – – – –
_____

Here is an apple.

Draw one more apple.

Now how many apples are there?
Write the number.

_____
– – – – – – – – – –
_____

Name _____

Cover the number **2** with glue.
Place dry macaroni on the glued area.
Let the glue dry.
Trace the number **2** with your finger.

Numbers 0—10 © 1988

Name _____

Trace the **2**s.

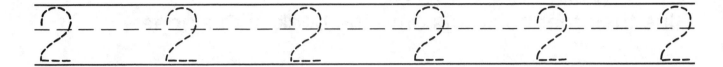

Write some **2**s.

_____
- - - - - - - - - - - - - - -
_____

Color **2** objects below.

Name _____

Color the tree and the coconuts. Cut out the coconuts.
Glue **2** coconuts to the tree.
Glue the other coconuts onto the back of this page.

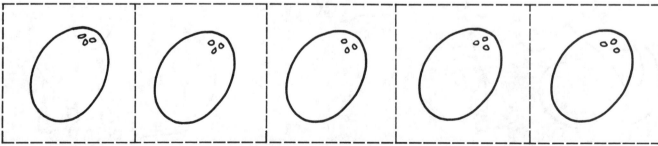

Numbers 0—10 © 1988

Name _____

Draw **2** ♡

Draw **2**

Draw **2**

Draw **2**

Name _____

Count the peanuts in each circle.
Cut out each circle that has **2** peanuts in it.
Glue those circles under the elephant's trunk.

Write the number **2**.

_____

— — — — — — — — — — — — — — — — — —

_____

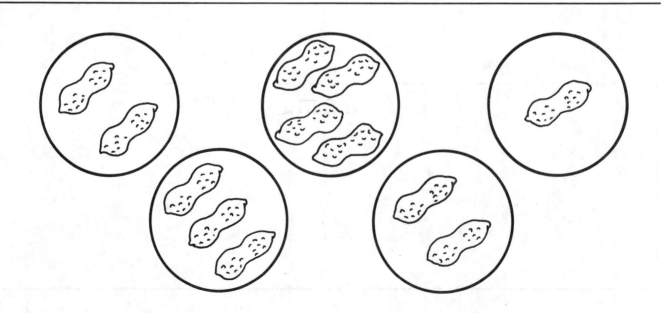

Numbers 0—10 © 1988

Name _____

Trace the number word.   Write the number word.

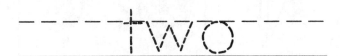

Draw a line from the number word to the correct number.

two

0 1 2 3 4 5

Draw a line from the number word to the correct number of objects.

two

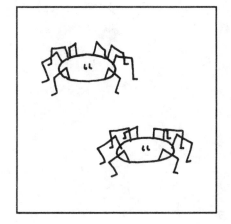
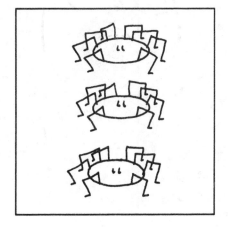
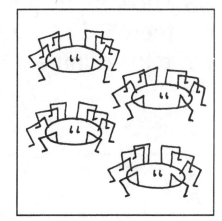

Name _____

Count the books.

Draw one more book.

Now how many books
are there?
Write the number.

_____
- - - - - - - - - - - -
_____

Count the flowers.

Draw one more flower.

Now how many flowers
are there?
Write the number.

_____
- - - - - - - - - - - -
_____

Numbers 0—10 © 1988

Name _____

Cover the number **3** with glue.
Place a piece of yarn on the glued area.
Let the glue dry.
Trace the number **3** with your finger.

Name _____

Trace the **3**s.

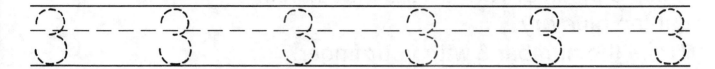

Write some **3**s.

_____

– – – – – – – – – – – – – – – – –

_____

Color **3** objects below.

Numbers 0—10 © 1988

Name _____

Color the coat and the buttons. Cut out the buttons.
Glue **3** buttons to the coat.
Glue the other buttons onto the back of this page.

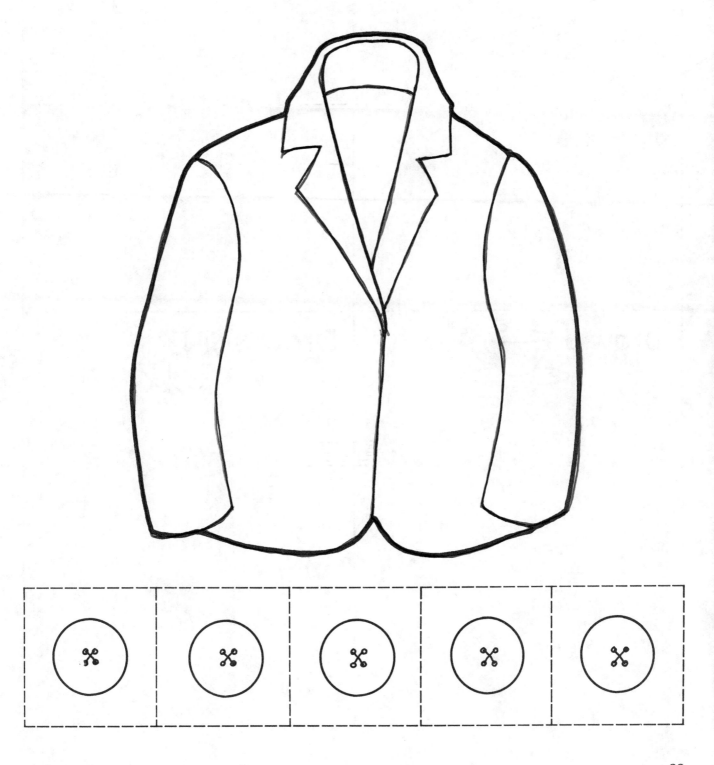

Name _____

Draw **3**

Draw **3**

Draw **3**

Draw **3**

Numbers 0—10 © 1988

Name _____

Color the igloo. Color the penguins below.
Cut out each group of **3** penguins.
Glue those groups beside your igloo.

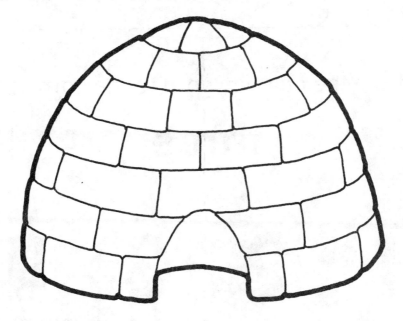

Write the number **3**.

Name _____

Trace the number word.   Write the number word.

---three--------   ------------------

Draw a line from the number word to the correct number.

three

0 1 2 3 4 5

Draw a line from the number word to the correct number of objects.

three

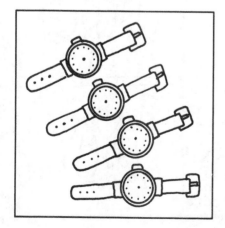   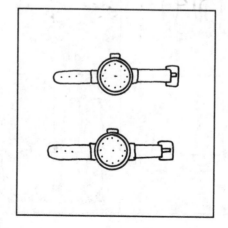   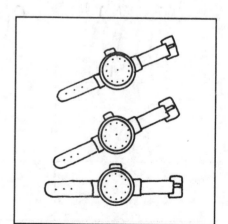

Name _____

Count the shirts.

Draw one more shirt.

Now how many shirts
are there?
Write the number.

_____

– – – – – – – – – –

_____

Count the frogs.

Draw one more frog.

Now how many frogs are
there?
Write the number.

_____

– – – – – – – – – –

_____

Name _____

Crumple pieces of tissue paper.
Cover the number **4** with glue.
Place the crumpled paper on the glued area.
Let the glue dry.
Trace the number **4** with your finger.

Numbers 0—10 © 1988

Name _____

Trace the **4**s.

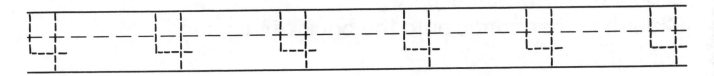

Write some **4**s.

_____

– – – – – – – – – – – – – – – – – – –

_____

Color **4** objects below.

Name _____

Color the dog and the bones. Cut out the bones.
Glue **4** bones beside the dog.
Glue the other bones onto the back of this page.

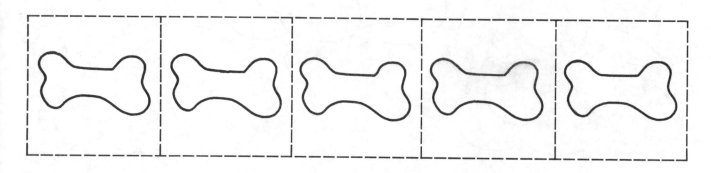

Numbers 0—10 © 1988

Name _____

Draw 4	Draw 4
Draw 4	Draw 4

Name _____

Count the arrows in each target.
Find the target that has **4** arrows in it.
Color the target.

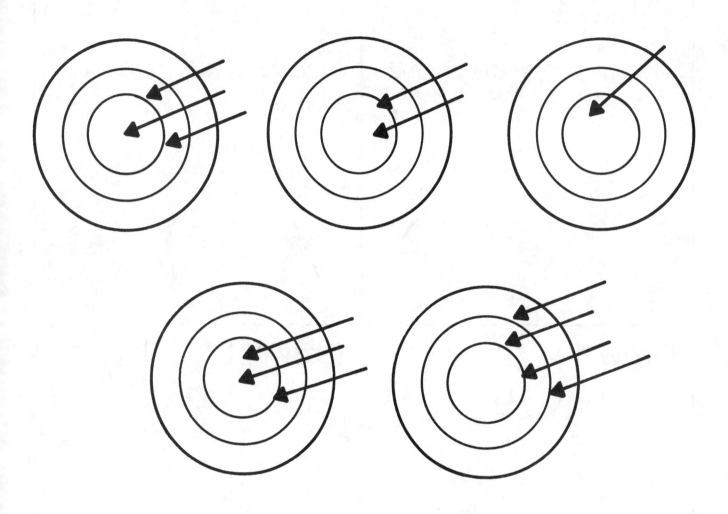

Write the number **4**.

_____

– –

_____

Name _____

Trace the number word. Write the number word.

 four

Draw a line from the number word to the correct number.

four

0 1 2 3 4 5

Draw a line from the number word to the correct number of objects.

four

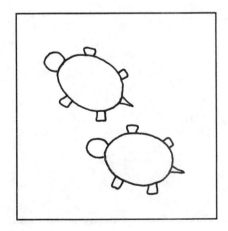  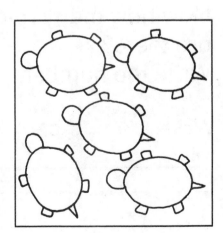

Name _____

Count the snails.

Draw one more snail.

Now how many snails
are there?
Write the number.

_____
_ _ _ _ _ _ _ _ _ _ _ _ _
_____

Count the rulers.

Draw one more ruler.

Now how many rulers
are there?
Write the number.

_____
_ _ _ _ _ _ _ _ _ _ _ _ _
_____

Numbers 0—10 © 1988

Name _____

Color the **5** using a black crayon.
Paint the whole page, brushing from left to right.
Let the paint dry.
Trace the number **5** with your finger.

Name _____

Trace the **5**s.

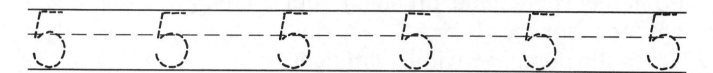

Write some **5**s.

_____

- - - - - - - - - - - - - - - - - -

_____

Color **5** objects below.

Numbers 0—10 © 1988

Name _____

Color the ball and the jacks. Cut out the jacks.
Glue **5** jacks beside the ball.
Glue the other jacks onto the back of this page.

Name _____

Draw **5**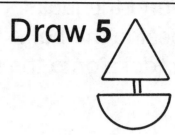

Draw **5**

Draw **5**

Draw **5**

Numbers 0—10 © 1988

Name _____

Color the logs and the beaver.
Count the logs in each group. Cut out the group of **5** logs.
Glue that group beside the beaver.

Write the number **5.**

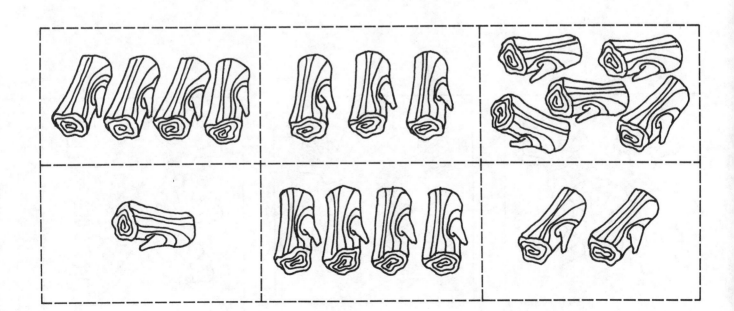

Name _____

Trace the number word.   Write the number word.

five

Draw a line from the number word to the correct number.

five

0 1 2 3 4 5

Draw a line from the number word to the correct number of objects.

five

Numbers 0—10 © 1988

Name _____

Count the ladders.

Draw one more ladder.

Now how many ladders
are there?
Write the number.

_____

– – – – – – – – – – – – –

_____

Count the fish.

Draw one more fish.

Now how many fish are
there?
Write the number.

_____

– – – – – – – – – – – – –

_____

Name _____

Glue **0** pieces of popcorn in the cloud that has the number **0**.
Glue **I** piece of popcorn in the cloud that has the number **I**.
Continue until all the clouds have the correct number of
popcorn pieces in them.

Numbers 0—10 © 1988

Name _____

Cut apart the circles below.
Glue **0** circles in the column marked **0**.
Glue **I** circle in the column marked **I**.
Continue until each column has the
correct number of circles.

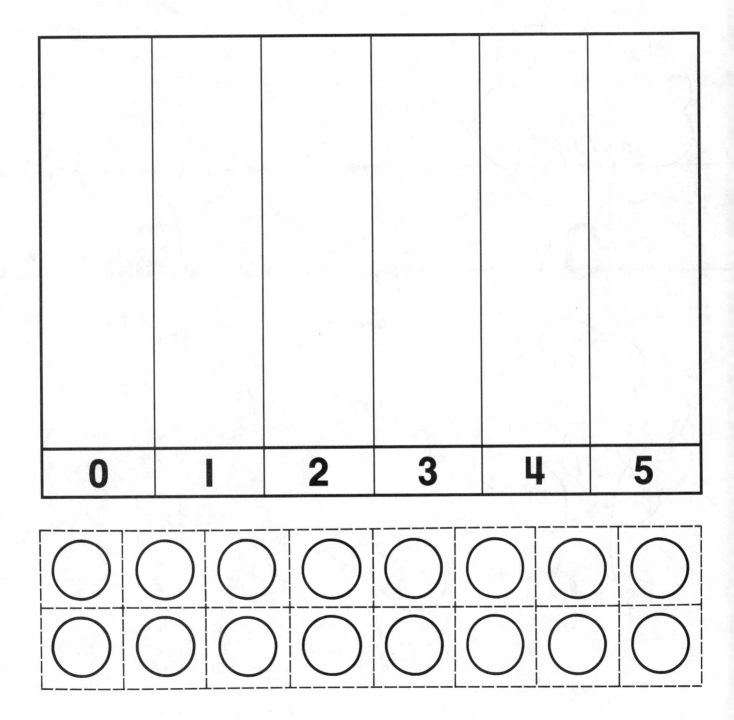

Name _____

Cut out each group of apples.
Count the apples in each group.
Glue each group of apples to the correct tree.
Color the picture.

Numbers 0—10 ® 1988

Name _____

Cover the number **6** with glue.
Place dried beans or peas on the glued area.
Let the glue dry.
Trace the number **6** with your finger.

Name _____

Trace the **6**s.

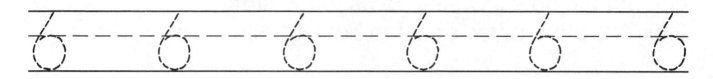

Write some **6**s.

Color **6** objects below.

Name _____

Color the musical notes below.
Cut out the notes. Glue **6** notes to the staff.
Glue the other notes onto the back of this page.

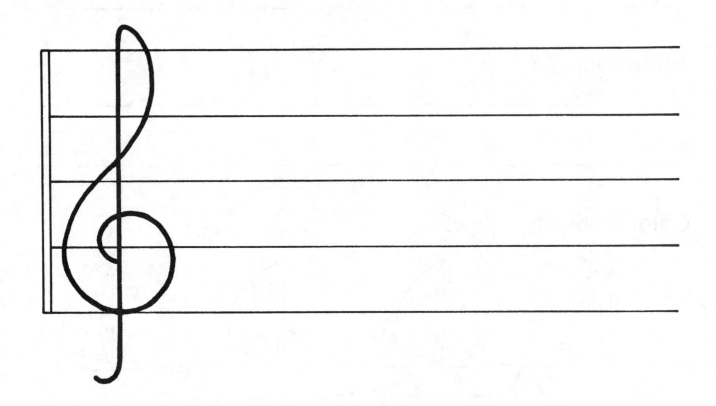

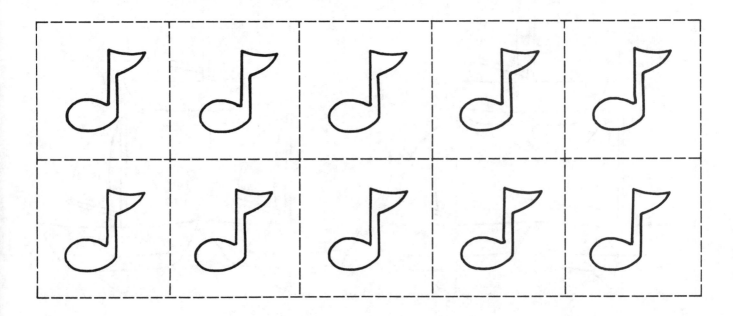

Numbers 0—10 © 1988

Name _____

Draw **6**

Draw **6**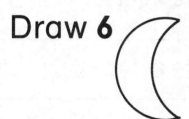

Draw **6**

Name _____

Color the spider and the webs.
Cut out the spider. Count the parts of each web.
Glue the spider inside the web that has **6** parts.

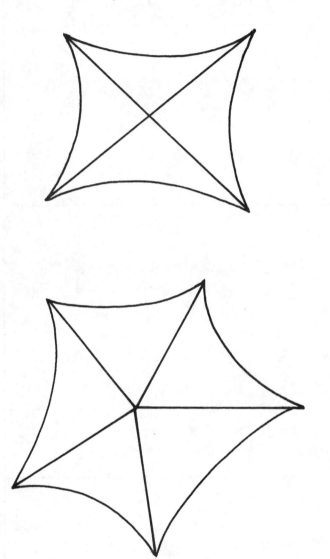

Write the number **6.**

_____

– – – – – – – – – – – – – –

_____

Numbers 0—10 © 1988

Name _____

Trace the number word.   Write the number word.

six

Draw a line from the number word to the correct number.

six

6 7 8 9 10

Draw a line from the number word to the correct number of objects.

six

Name _____

Count the hats.

Draw one more hat.

Now how many hats are there?
Write the number.

_____

— — — — — — — — — — — —

_____

Count the hands.

Draw one more hand.

Now how many hands are there?
Write the number.

_____

— — — — — — — — — — — —

_____

Numbers 0—10 © 1988

Name _____

Cover the number **7** with glue.
Sprinkle sand on the glued area.
Let the glue dry.
Trace the number **7** with your finger.

Name _____

Trace the **7**s.

7   7   7   7   7   7

Write some **7**s.

Color **7** objects below.

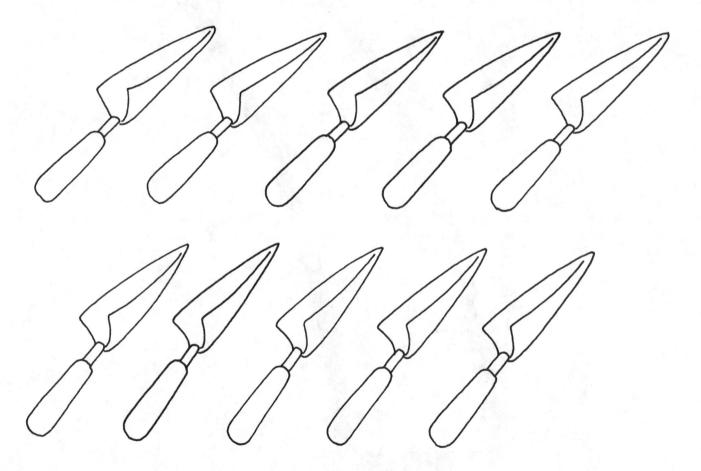

Numbers 0—10 © 1988

Name _____

Color the birdfeeder and the birds. Cut out the birds.
Glue **7** birds to the birdfeeder.
Glue the other birds onto the back of this page.

Name _____

Draw **7**

Draw **7**

Draw **7**

Numbers 0—10 © 1988

Name _____

Count the chicks in each circle.
Find each group of **7** chicks.
Color the chicks in each group of **7**.

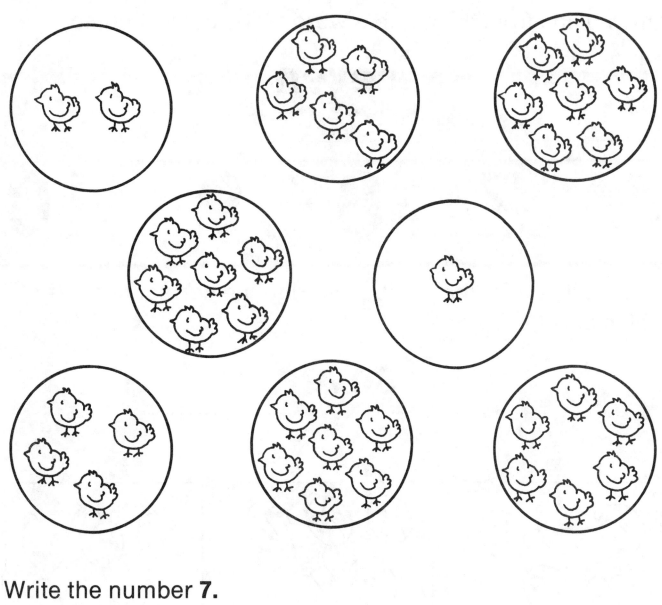

Write the number **7**.

_____

– –

_____

Name _____

Trace the number word.　　　　Write the number word.

‾‾‾‾‾‾‾‾‾‾‾‾‾‾‾‾‾‾‾‾‾　　　　‾‾‾‾‾‾‾‾‾‾‾‾‾‾‾‾‾

seven　　　　- - - - - - - - - - -

Draw a line from the number word to the correct number.

seven

6　　7　　8　　9　　10

Draw a line from the number word to the correct number of objects.

seven

　　　　Numbers 0—10 © 1988

Name _____

Count the pencils.

Draw one more pencil.

Now how many pencils
are there?
Write the number.

_____
_ _ _ _ _ _ _ _ _ _ _
_____

Count the cactuses.

Draw one more cactus.

Now how many cactuses
are there?

Write the number.

_____
_ _ _ _ _ _ _ _ _ _ _
_____

Name _____

Find the number **8** in newspapers or magazines.
Cut some out.
Glue them inside the number **8** below.
Let the glue dry.
Trace the number **8** with your finger.

Numbers 0—10 © 1988

Name _____

Trace the **8**s.

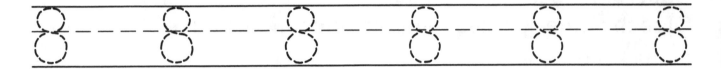

Write some **8**s.

_____

- - - - - - - - - - - - - - - - - - -

_____

Color **8** objects below.

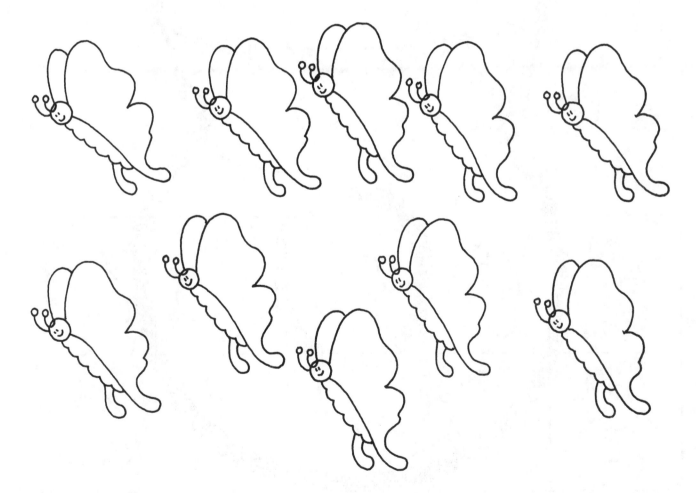

Name _____

Color the basket and the berries.
Cut out the berries. Glue **8** berries inside the basket.
Glue the other berries onto the back of this page.

Numbers 0—10 © 1988

Name _____

Draw **8**

Draw **8**

Draw **8**

Name _____

Count the dots on each hat.
Color the hat that has **8** dots.

Write the number **8.**

_____

– –

_____

Numbers 0—10 © 1988

Name _____

Trace the number word.

Write the number word.

Draw a line from the number word to the correct number.

eight

6 7 8 9 10

Draw a line from the number word to the correct number of objects.

eight

Name _____

Count the pears.

Draw one more pear.

Now how many pears
are there?
Write the number.

_____

- - - - - - - - - - - -

_____

Count the stars.

Draw one more star.

Now how many stars are
there?
Write the number.

_____

- - - - - - - - - - - -

_____

Numbers 0—10 © 1988

Name _____

Glue popcorn inside the number **9**.
Let the glue dry.
Trace the number **9** with your finger.

Name _____

Trace the **9**s.

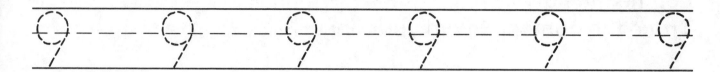

Write the **9**s.

Color **9** objects below.

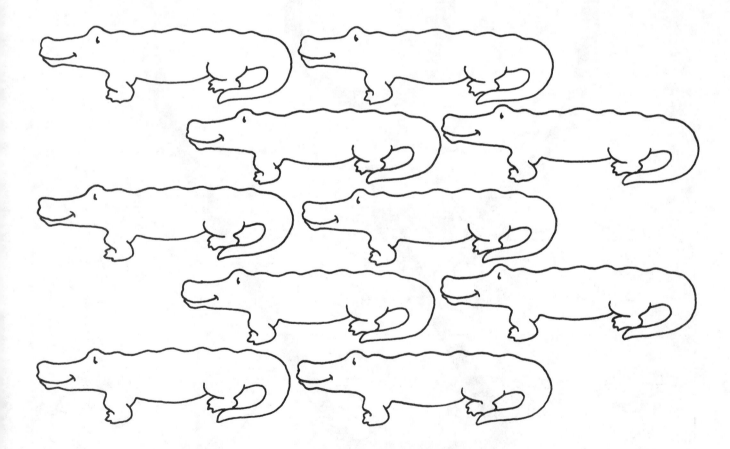

Numbers 0—10 © 1988

Name _____

Color the tree and the pinecones.
Cut out the pinecones. Glue **9** pinecones on the tree.
Glue the other pinecones onto the back of this page.

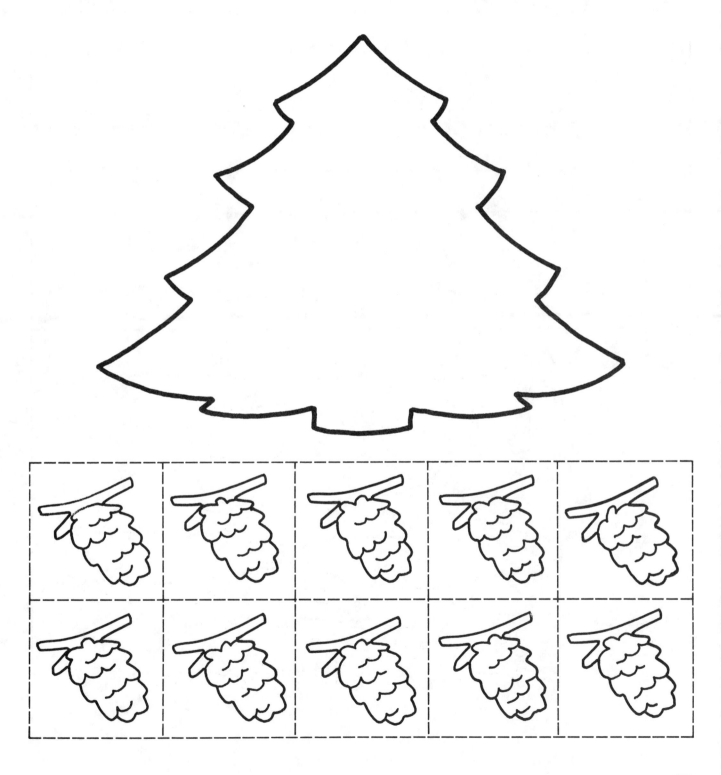

Name _____

Draw **9**

Draw **9**

Numbers 0—10 © 1988

Name _____

Count the carrots in each bunch.
Color each bunch that has **9** carrots in it.

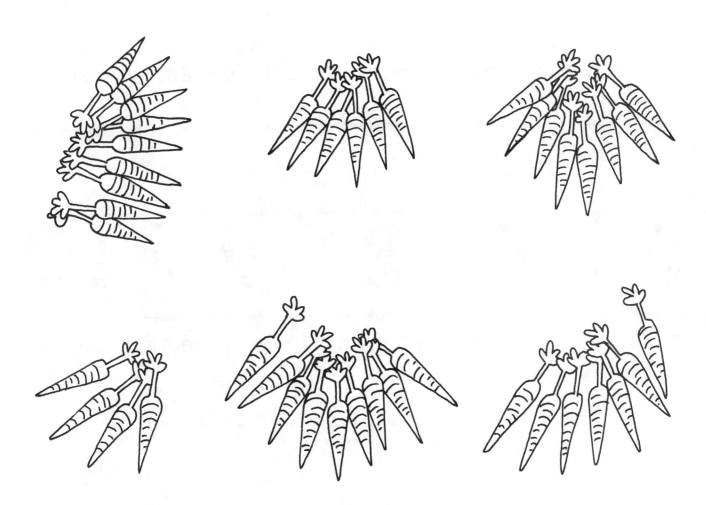

Write the number **9.**

_____

– –

_____

Name _____

Trace the number word.   Write the number word.

nine

Draw a line from the number word to the correct number.

nine

6   7   8   9   10

Draw a line from the number word to the correct number of objects.

nine

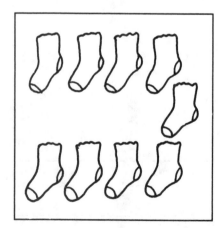

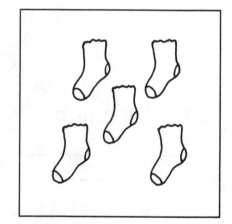

Numbers 0—10 © 1988

Name _____

Count the triangles.

Draw one more triangle.

Now how many triangles
are there?
Write the number.

Count the fruit juice bars.

Draw one more fruit
juice bar.

Now how many fruit juice
bars are there?
Write the number.

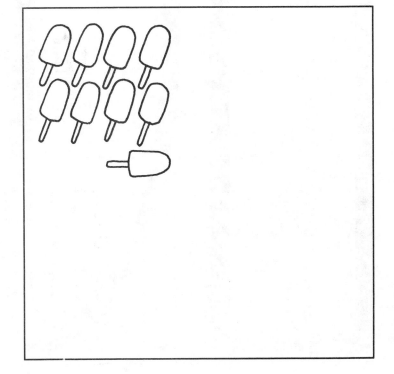

Name _____

Color the **10** using a black crayon.
Paint the whole page, brushing from left to right.
Let the paint dry.
Trace the number **10** with your finger.

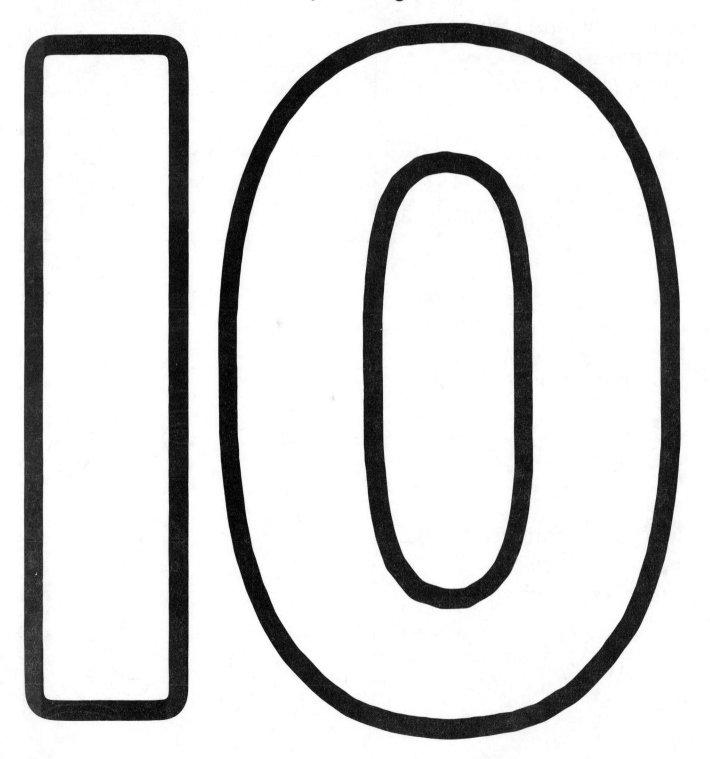

Numbers 0—10 © 1988

Name _____

Trace the **10**s.

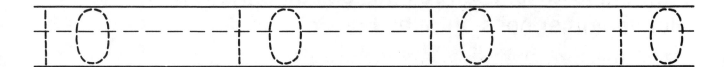

Write some **10**s.

_____

— — — — — — — — — — — — — — —

_____

Color **10** objects below.

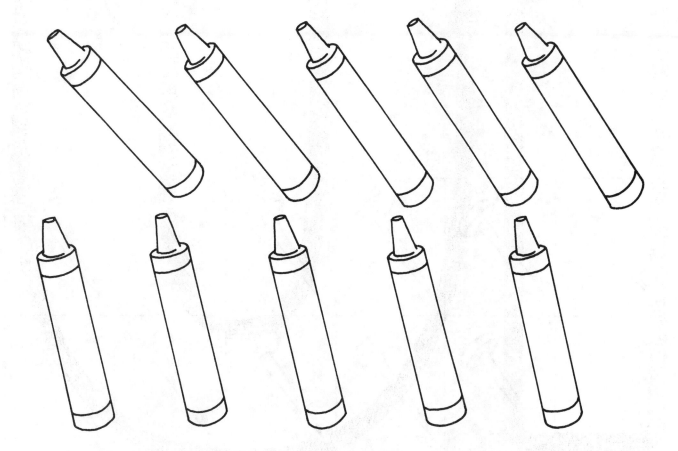

Name _____

Color the hammer and the nails.
Cut out the nails. Glue **10** nails beside the hammer.
Glue the other nails onto the back of this page.

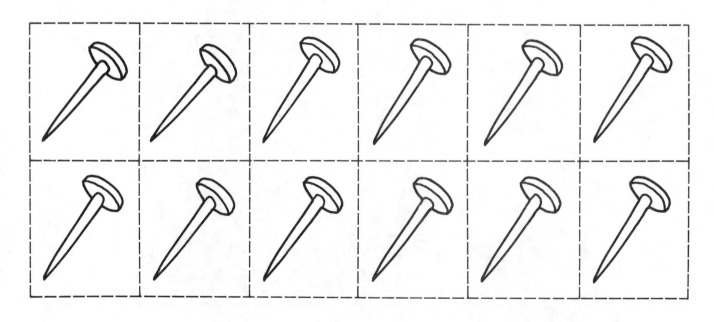

Numbers 0—10 © 1988

Name _____

Draw **10**

Draw **10**

Name _____

Count the puffs of steam above each kettle.
Find the kettle that has **10** puffs of steam.
Color that kettle. Draw a teacup beside the kettle.

Write the number **10.**

_____

– –

_____

Numbers 0—10 © 1988

Name _____

Trace the number word.   Write the number word.

ten

Draw a line from the number word to the correct number.

ten

6   7   8   9   10

Draw a line from the number word to the correct number of objects.

ten

 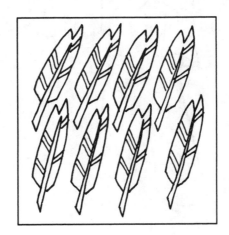 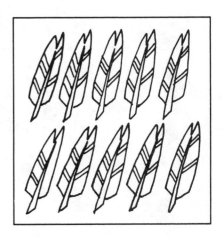

Name _____

Glue **6** sequins in the balloon that has the number 6.
Glue **7** sequins in the balloon that has the number 7.
Continue until all the balloons have the correct number of
sequins in them.

Numbers 0—10 © 1988

Name _____

Cut apart the rectangles.

Glue **6** rectangles in the column marked **6**.

Glue **7** rectangles in the column marked **7**.

Continue until each column has the correct number of rectangles.

6	7	8	9	10

Numbers 0—10 © 1988

Name _____

Cut out each group of honeybees. Count the bees in each group. Glue each group of bees below the correct bear. Color the picture.

Numbers 0—10 © 1988

Name _____

Glue **0** beans in the nest marked **0**.
Glue **1** bean in the nest marked **1**.
Continue until all the nests have the correct number of beans in them. What do the beans represent?

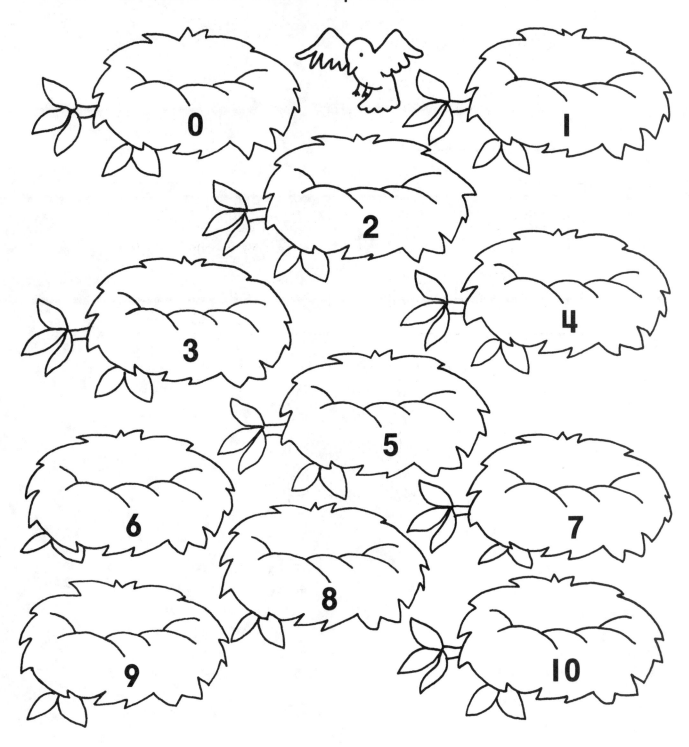